WORLD'S WORST...

CHEMICAL Disasters

Rob Alcraft

Heinemann Library
Chicago, Illinois

Customer Service 888-454-2279

Designed by Celia Floyd
Illustrations by David Cuzik (Pennant Illustration) and Jeff Edwards
Originated by Dot Gradations
Printed by Wing King Tong, in Hong Kong

04 03 02 01 00
10 9 8 7 6 5 4 3 2 1

Library of Congress Cataloging-in-Publication Data

Alcraft, Rob, 1966-
  Chemical disasters / Rob Alcraft.
    p. cm. – (World's worst)
  Includes bibliographical references and index.
  Summary: Examines the events leading to the poisoning of Minamata
Bay, the nitrate explosion in Texas City, and the gas cloud
disaster in Bhopal, the consequences of the accidents, and how
  they might have been averted.
  ISBN 1-57572-987-3 (library binding)
  1. Chemical industry—Accidents Juvenile literature.
 [1. Chemical industry—Accidents.] I. Title. II. Series.
 TP150.A23A38 1999
 363.17'91—dc21   99-34942
                                                            CIP

Acknowledgments

The Publishers would like to thank the following for permission to reproduce photographs:
Tony Stone/Wayne Eastep, p. 2; Science Photo Library/Perquis, p. 5; Science Photo Library/Garry Watson, p. 6; Ann Ronan, p. 7; Rex Features/SIPA, pp. 9, 21; Corbis-Bettman/Yamashita, p. 12; Hulton Getty, p. 14; Corbis-Bettman, pp. 15, 19; Heinemann, p. 18; Corbis-Bettman/Rainier, p. 24; Tony Stone/Keith Wood, p. 26; Tony Stone/David Woodfall, p. 27; Telegraph Colour Library/Bora Merdsoy, p. 29.

Cover photograph reproduced with permission of Tony Stone/David Woodfall.

Every effort has been made to contact copyright holders of any material reproduced in this book. Any omissions will be rectified in subsequent printings if notice is given to the Publisher.

Some words are shown in bold, **like this**. You can find out what they mean by looking in the glossary

Contents

A Chemical World

We live in a world of chemicals. There are over 60,000 artificial chemicals in everyday use. All sorts of plastics, clothes, and food are made and colored with chemicals. Chemicals make the medicines that make us better. Chemicals keep our homes clean. Farmers use them as **fertilizer** and to kill pests and weeds.

When chemicals are mixed or heated, they make new substances. Chemical factories make and mix chemicals to make acids and dyes that are used in many industries. The most common **raw materials** for making chemicals are oil and coal.

Farmers spray fields of crops with chemical **pesticides** and fertilizers to produce larger harvests. These chemicals can be harmful to the **environment**.

Like many chemicals, the **synthetic** rubber in these rubber gloves is made from oil, gas, and coal.

Many countries have laws that control how chemicals are made. The **World Health Organization**—an international organization that is part of the **United Nations**—checks the safety of chemical industries around the world. It also monitors accidents and spills. Chemical factories have to be safe. Workers are trained so they know what to do in an emergency. Many countries have inspectors who check chemical factories.

Acids and alkalis

There are two families of chemicals—acids and alkalis. Lemons contain acid. That's why they have a sharp, sour taste. Strong acids are very dangerous, even to touch. They burn the skin and can eat through wood, cloth, and metal. Acids are used in many products, including batteries, explosives, and fertilizers.

Alkalis are soapy to the touch. Strong alkalis will burn your skin. Alkalis are used in soap and to make glass.

Handle with Care

Chemicals are useful, but they can be dangerous. Many are poisonous. Many cause diseases and illness. Many can burn. Many are highly explosive. If they are not used properly, they can kill.

Because chemicals are dangerous, even small accidents can be disastrous. Very small amounts of some chemicals can be harmful if released into air or water. Chemicals are often invisible. This makes them hard to detect. If they leak into the **environment**, they can build up over many years. The effects of a chemical disaster can take many years to show up. The first sign can be when people become ill, or a river or forest dies.

The skull and crossbones symbol means danger—in this case, chemical danger.

Dangerous chemicals

Your body—a chemical factory

Our bodies contain chemicals. The chemicals react and work together. They are controlled by **enzymes**, which keep our bodies working properly. Your body contains around 30,000 enzymes, making it one of the most complicated chemical factories around.

In this book, we will look at three of the world's worst chemical disasters. These disasters shocked the world. In many ways, they changed the view of safety in the chemical industry. We will look at what happened. What went wrong? Have we learned how to avoid disasters?

This drawing of the Mad Hatter, from Lewis Carroll's *Alice's Adventures in Wonderland,* is by John Tenniel.

Mad as a Hatter?

Remember the Mad Hatter from *Alice's Adventures in Wonderland*? In the past, hat makers really did go mad. They used materials treated with chemicals that contained poisonous mercury. After years of breathing in mercury **fumes**, the nervous systems of many hat makers were affected, and they would shake and slur their words.

Poison!
The Poisoning of Minamata Bay

Japan

Minamata Bay

In 1932, a plastics factory in Minamata Bay, Japan, began dumping its **toxic** mercury waste into the sea. It was the beginning of a pollution disaster that would kill hundreds of people and disable thousands more.

Death by pollution

First, the fish began to die. Dead fish littered Minamata Bay's shining blue waters and washed up on the beaches. But no one knew why. Next, cats and crows began to die.

Then, in 1953, people complained of strange illnesses. They were slurring their speech and feeling dizzy. People began to die. When doctors examined the bodies, they found that the victims' brains had been damaged. They had no idea what caused the damage. They named the mysterious illness Minamata Disease.

As the deaths continued, only a company named Chisso knew the facts. The people of Minamata Bay were being poisoned by mercury waste from the Chisso plastics factory. In 1956, the Chisso Company doctor examined cats that had drunk waste water from the factory. He discovered that the mercury in the water had killed them. But executives at the factory ordered the cats and the evidence to be destroyed. The Chisso Company continued to pump the mercury waste into the waters of Minamata Bay.

Many of the victims of the Minamata Bay disaster were unborn children. They were poisoned in the womb, from food eaten by their mothers. Mrs. Sakamoto's daughter, Shinobu, was poisoned by mercury

We took her to the hospital. By then the cause of the poisoning was known. I was told there was no cure. We could only try to train her to use the undamaged parts of her body and brain.

At first our only hope was that she could walk. Then, we prayed that she could go to school. Now our hope is that she will be able to take care of herself when we no longer can.

A poisonous metal

Mercury is a potentially lethal poison. It can cause illness, madness, and eventually death. It gets into the body through the skin, through **fumes**, or through **contaminated** food. In the Minamata Bay disaster, fish and shellfish in the bay absorbed the mercury from the factory waste. Fishermen and their families became ill when they ate the fish.

Contaminated water and waste gushed freely into Minamata Bay.

The Making of a Disaster

The Chisso Company knew that the mercury it was dumping into Minamata Bay was killing sealife and the people and animals that fed on it. Yet the company dumped mercury-rich water into the bay until 1968.

The disaster was slow and lingering. More than 12,000 people were affected, and over 900 people died. Children were born sick and disabled.

1. In 1932, the Chisso Company begins releasing mercury into Minamata Bay.

2. Fish begin to die in the 1950s. Tests on crabs and fish in the bay find that they are **contaminated** with large amounts of mercury. Cats and birds that eat the fish from the polluted bay start dying.

3. **Bacteria** in Minamata Bay change the mercury waste into methyl mercury. Methyl mercury is even more deadly than mercury itself.

4. In 1956, after people and animals die, a doctor at the Chisso Company traces the causes of Minamata Disease to mercury in Chisso's own factory waste. Nothing is done. The Japan Chemical Industry Association helps Chisso by supplying scientists to find other reasons for Minamata Disease—reasons that do not blame Chisso.

7. The town of Minamata is scarred by the disaster. Thousands of people leave. A population of 50,000 shrinks to 32,000. The fishing industry, which many people rely on, is destroyed.

6. In 1966, Chisso finally stops dumping mercury waste in Minamata Bay. The national government does not officially blame Chisso for the disaster until 1968. The Japanese government never accepts responsibility.

5. Later in 1956, the Japanese government officially recognizes the link between eating fish from Minamata Bay and Minamata Disease. Again, nothing is done.

The 65-Year Clean-up

Reaction to the Minamata disaster was slow. It took the Japanese government eleven years to blame Chisso for the pollution. It took another twenty years before trials and appeals in the courts met with any success. It wasn't until 1995 that Japan's Supreme Court found the President of the Chisso Company and the director of the Chisso Minamata factory guilty of causing the disaster.

Many victims of the Minamata Bay disaster suffered terrible disabilities. These survivors are lucky to be alive to enjoy this outing to see the spring cherry blossoms.

A company town

The Chisso Company factory provided the town of Minamata with most of its jobs. One third of all the employed people in the town worked there. Nearly two-thirds of local taxes came from the factory. Most of Minamata's council and mayors had once been workers or managers at the Chisso factory. Because of this, no one wanted to criticize the Chisso Company—even if it was suspected of poisoning Minamata Bay.

Blame was very important in the Minamata disaster. Managers at Chisso had to take responsibility for what they had done. Chisso had to pay **compensation** to those who had suffered and pay to clean the poisoned sea. In 1995, Chisso agreed to pay out 4.94 billion Japanese yen (over $50 million) to five groups of patients. Yet not everyone has been compensated. Although 12,615 people are Minamata Disease sufferers, only 3,000 have been officially recognized and paid compensation. It seems that many victims will never get the justice they deserve.

Cleaning Minamata Bay

The cleaning of Minamata Bay took many years. In 1974, a 1-mile (2-kilometer) net was stretched across the bay to trap fish. This kept poisoned fish from finding their way onto people's plates. In 1977, poisoned sludge was removed. Then, in 1983, a poisonous mud was **dredged** from the bay. If the mud was removed, the **environment** would recover. People would be safe. In 1997, fish from Minamata Bay were declared safe to eat. The nets were taken away—65 years after the pollution began.

Knowing and complaining

As the people of Japan learned about Minamata Disease, they became worried about pollution in their own areas. In 1960, the Environment Agency of the **OECD** found that very few complaints were made to the authorities about pollution. By 1972, over 86,000 complaints were made. People were realizing the dangers of pollution—and wanted something done about it!

Explosion!

The Destruction of Texas City

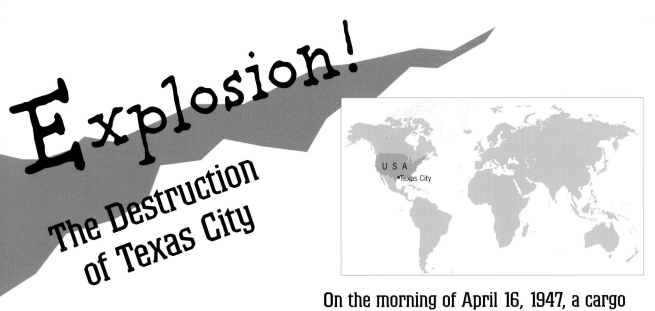

USA
•Texas City

On the morning of April 16, 1947, a cargo ship in Texas City harbor, Texas, caught fire and exploded. Over 500 people died. More than 50 years later, the people of Texas City still remember what they call "the explosion."

No Ordinary Day

It was a cool, clear morning. The Texas City harbor was busy with ships. It seemed like any ordinary day. But one ship would change everything. Shortly before 9:00 A.M., the *Grandcamp*, bound for Europe, caught fire. Thick black smoke billowed out across the water.

In the explosion at Texas City Harbor, a huge steel barge was blown from the ship basin to dry land, destroying several cars.

A crowd gathered to watch. Almost thirty firefighters from the Texas City Volunteer Fire Department arrived to tackle the blaze. As the smoke shifted in the breeze, people saw it was a strange orange color. This was the first sign of disaster.

The *Grandcamp* was carrying nitrate **fertilizer**, which is highly explosive. Suddenly, a little after 9 A.M., the *Grandcamp* exploded. Smoke shot 1,890 feet (600 meters) into the air. The blast smashed across the harbor, sending a tidal wave into the town. Entire buildings collapsed in flames. A chemical plant, warehouses, and **refineries** began to burn. Texas City was going up in flames.

I had just gotten out of bed when I felt the house shaking to beat the band. I thought I saw bodies wheeling through a whole bunch of debris in the air. [Then I saw] a dark ridge (the shock wave from the blast) moving 3 meters [10 feet] off the ground, heading toward the house.

John Hill, a chemical engineer working at a plant in Texas City on the morning of the explosion.

This chemical plant was one of the buildings devastated by the Texas City blaze.

Texas City Just Blew Up!

Fires started everywhere. All day, Texas City burned. Firefighters and rescuers struggled to find and treat survivors, and to fight the fires that raged throughout the city. Frank Simpson, a student outside the city, was told, "Texas City just blew up."

But the disaster wasn't over. The *High Flyer*, another ship out in the wrecked harbor, was burning. On board was more explosive nitrate. Tugboats struggled to tow the *High Flyer* out into the sea, but it was stuck on wreckage from the first explosion. Night came and the ship burned on. At 1:10 A.M., the *High Flyer*—still in the harbor—finally exploded.

1. Warehouses and **refineries** border the harbor. The *Grandcamp*, bound for Europe, is loaded with nitrate **fertilizer**. When the fire starts, no one knows its cargo is dangerous. Crowds gather to watch the fire.

2. The *Grandcamp*'s steam fire control system is switched on. But instead of smothering the fire, the heat and pressure react with the nitrate cargo and produce explosive gas.

3. *Grandcamp*'s entire cargo explodes. The blast wrecks the harbor, hurling debris into the air. A nearby chemical plant burns. More explosions threaten, but almost all the fire equipment in Texas City is destroyed in the first blast. Many of the volunteer firefighters at the scene are killed.

16

6. Fire burns through Texas City for a week after the disaster. A month goes by before the last body is pulled clear of the collapsed buildings. Many of the dead are never identified.

5. In the harbor, the *High Flyer* is on fire. It is also carrying an explosive cargo of nitrate. Tugboats try to pull it clear of the harbor, but fail. At 1:00 A.M., people are ordered to clear the area. At 1:10 A.M., the ship explodes. This explosion is larger and more devastating than the first.

4. Beyond the harbor, a wave of water from the blast crashes into the town. Buildings crumble in the blast and begin to burn. The *Grandcamp*'s 1.5-ton anchor is blasted into an oil refinery almost 2 miles (3 kilometers) away. The explosion is heard 145 miles (240 kilometers) away.

Getting Justice

Fires and explosions had become a normal part of life in Texas City. It was a booming industrial town. The harbor was bordered by **refineries**, warehouses, and chemical plants. When oil tanks or warehouses went up in flames, people would flock to see them. It was public entertainment. Safety wasn't given the priority it needed. All this changeed with the explosion of the *Grandcamp* and the disastrous fires that followed. As a result of the devastating disaster, Community Awareness Response in Emergency teams were set up to warn people of dangerous situations.

This fountain was built as a memorial to the fire fighters who died in the Texas City blaze.

Without warning

When the *Grandcamp* caught fire, the port authorities should have warned the firefighters what was on board and what the danger was. There should have been an **evacuation**. Instead, schoolchildren and dock workers watched as the explosive ship burned. Ken LeMat was president of the Texas City Company Railway, which ran the port. He recognized that people should have done more. "We weren't prepared for the disaster," he said. "We lacked communications and there was no one to take control."

These survivors of the explosion were lucky to escape unharmed. The port still holds terrible memories for many people.

Compensation

Victims of the Texas City disaster tried to sue the U.S. Coast Guard. They accused them of **negligence**. The believed that the Coast Guard should have done more to prevent the disaster. But the U.S. Supreme Court ruled that the Coast Guard, as a government agency, couldn't be held responsible under existing U.S. laws.

It was not until 1955—eight years after the disaster—that Congress passed a new law and finally paid **compensation** to the victims who had suffered in the Texas City disaster.

Killer Cloud!
Gas Cloud Disaster in Bhopal

At midnight on December 2, 1984, a deadly cloud of gas leaked from the American-owned Union Carbide **pesticide** factory in the city of Bhopal, India. Within a week, 2,500 people were dead. It was disaster on a tragic scale. People are still suffering and dying from the effects of the poison.

An accident waiting to happen

A new underground tank, called tank 610, had been installed at the Bhopal factory. It held 42 tons of a dangerous liquid chemical called methyl isocyanate. This is a volatile chemical that reacts violently if it is heated or mixed with air or water.

On December 1, engineers at the plant were cleaning pipes above tank 610. They flushed thousands of gallons of water through the system. The routine procedure should have been safe. But the engineers didn't know that a **valve** was leaking. Water that should have flushed through the pipes had in fact leaked into tank 610. Inside the tank, the water and volatile chemicals began to mix and react dangerously together.

Later, workers noticed water leaking from tank 610. They found that a **pressure gauge** was missing. They guessed it had been blown off by the pressure building up in the tank. At midnight, they told the factory boss that gas was leaking into the air.

20

In the factory control room, the emergency had become clear. The temperature and pressure in the tank were off the scale. Tank 610 was boiling like a kettle, letting off a cloud of poison gas. Workers at the factory were **evacuated**. But for the city of Bhopal there was no warning. A fog of deadly yellow gas began to drift slowly over the city.

After the Bhopal disaster, street clinics were set up to treat victims.

Ramesh was a young boy when disaster struck Bhopal. This is his story

People were shouting, "Get up, run, run. Gas has leaked!" My elder brother got up and said, "Everyone is running away. We must run too." I opened my eyes and saw the room was full of white smoke.

The moment I took the rug from my face, my eyes started stinging and every breath was burning my insides. I was scared of opening my eyes.

As we reached the main road we could see a lot of people lying around. We did not know whether they were dead or unconscious.

Unfolding Disaster

For three hours, the gas leaked over Bhopal. People woke to find their houses filled with choking, blinding gas. They did not know what the gas was. Union Carbide, which owned the factory, had never told people what to do in an emergency.

The poison cloud affected 200,000 people in Bhopal. In the darkness there was panic. People ran through the streets trying to escape. Thousands of people crammed into hospitals. Many could not breathe. Many had been blinded by the methyl isocyanate gas. By morning, hospitals were overflowing. There weren't enough hospitals to treat everyone who needed help.

1. Inspectors from Union Carbide visit Bhopal in 1982. They note that the Union Carbide factory is unsafe and would not be allowed in the United States. There are accidents the next year. A boy dies, but the factory is not made safer.

2. Sometime late in the evening of December 2, 1984, water leaks into tank 610. Temperature and pressure in the tank rise as the water and chemical inside react and begin to boil. The chemical turns into gas and begins to escape from the tank.

3. Two emergency systems should keep gas from escaping from tank 610. The first is a filter called a scrubber. It should make the deadly gas harmless, but it doesn't work. Next, a tower should burn gas as it escapes, but the tower is closed for repair.

5. By the early hours of December 3, hospitals are overflowing with injured and dying people. There are not enough drugs or doctors to help everyone. Only the most seriously hurt get treatment. No one is quite sure what the gas contained or how to treat the injuries. Union Carbide has information about the gas that can help treat the injured but does not tell doctors or authorities at Bhopal. More people die. Even after 2,000 people are dead, Union Carbide factory managers are still claiming that the gas was not poisonous.

4. By midnight on December 2, the gas cloud is drifting across Bhopal. People don't know what to do or where to go. When warning sirens sound, the panic and chaos increase.

Shock and Anger

In the hours and days after the gas leak, the people of Bhopal were in a state of shock. Parents and children just wanted to find their families and bury their dead.

But this shock quickly turned to anger. Indian people realized that this disaster would never have been allowed to happen in the United States.

People became so angry that workers who had been at the factory on the night of the disaster had to be given protection. When Warren Anderson, the chairman of Union Carbide, arrived from the United States, he was arrested by police in Bhopal. He was released later, but it was clear that in Bhopal people wouldn't forget what had happened.

Living with the disaster

Many of those who survived in Bhopal suffered diseases and injuries after the disaster. Many were blinded. Other people's lungs were damaged by breathing the poisonous gas. Yet many people never got the help they needed. Money to help survivors and the families of those who had died took months or years to distribute. It took five years before Union Carbide agreed to pay anything to the surviving victims. The amount they paid was much less than a U.S. court would have made the company pay if the disaster had happened in the United States.

Angry protesters demonstrate against Union Carbide.

The disaster continues to affect the survivors of Bhopal. Between 1,000 and 10,000 people have died from the effects of the gas since the disaster. But Union Carbide has tried to open the factory again. Even today, the company denies fault for the disaster. The company's position is this: one worker **sabotaged** the plant on purpose.

Taking care

Disasters often make people realize they should be more careful. This is what many governments and chemical companies realized after the disaster at Bhopal, India. Since 1986, all U.S. chemical companies have been required to report all leaks and spills of any harmful chemicals. There is a long list of these chemicals. Many chemical companies also spend more money on safety and the **environment**. Today, they spend around four percent of their total sales on safety and the environment. This is four times what they spent before the Bhopal disaster.

Safety and Money

Safety and pollution control—both cost money. At Bhopal and Minamata, companies decided to save money. These decisions cost the lives of thousands of people.

Organizations such as **Greenpeace** say that industries should take precautions, not risks. They call this the precautionary principle. Yet this would mean that the things we buy would become more expensive. Companies that spend money on safety would try to get that back from their customers. Not everyone wants or is able to pay more.

Industries make and use some of the most dangerous substances we have ever known. Sometimes the dangers aren't realized until these chemicals have been used for years. This is true of a **pesticide** known as DDT, which was used all over the world for decades. It is now known that DDT harms people and causes disease. Today, DDT is banned by many of the world's governments.

This foul-looking fluid is chemically polluted water gushing out from an industrial waste pipe in Merseyside, England.

Technicians wearing protective suits take soil samples at a toxic waste area. They can assess how much damage is being done to the environment.

Finding a balance

Governments have the job of finding a balance between cost and safety. They make rules about how much a factory can pollute and what safety systems it should have in order to avoid disaster. But, as science and industry move farther ahead into the unknown, disasters can happen.

An everyday disaster

Pollution from industry and trash is an everyday disaster. Although most companies obey strict guidelines, some chemicals end up in the **environment**. From there they find their way into our food and drinking water. Other chemicals are added to our food when it is made or processed. Some cause harm, but some do not. It is not yet known which are the really dangerous ones.

Waste sites

Some hazardous waste cannot be destroyed—it must be stored. Sometimes it is not stored safely. In the United States, there are over 425,000 potentially hazardous waste sites—one for every 611 Americans.

The World's Worst Chemical Disasters

Explosion—Texas City, Texas, April 16, 1947 An explosion on a ship carrying dangerous chemicals in Galveston Bay costs hundreds of lives. Fires and explosions rage through the city, flattening buildings.

Pollution—Minamata Bay, Japan, 1953–1968 Mercury waste dumped by a plastics factory kills 900 people.

Blast—Flixborough, England, June 1, 1974 A chemical plant explodes, killing 55 people and injuring 75.

Poison cloud—Seveso, Italy, July 10, 1976 Poisonous dioxin escapes from a chemical factory. Over 700 people are **evacuated**, but 250 people are poisoned. One year later over 400 children become ill.

Poison cloud—Bhopal, India, December 2, 1984 Highly **toxic** gas leaks from the Union Carbide **pesticide** factory. 2,500 people are dead within a week.

Pollution—Basel, Switzerland, November 1, 1986 Thirty tons of chemicals are flushed into the Rhine River after a fire at a chemical plant. The river takes ten years to recover.

Global warming

It is hard to know what disasters we might encounter in the future. Global warming—a predicted rise in temperatures around the world—could be one of them. Some scientists say it is already happening. Between 1890 and 1990, the world has become 1°F (0.5°C) warmer.

Global warming is caused by gases, such as carbon dioxide from car exhausts and industry, that collect high in the earth's **atmosphere**. Instead of heat from the sun hitting the earth and being bounced out again, these gases trap the heat and return it to Earth. As a result, the world becomes warmer.

If you live in a cold country, the idea of being a little warmer might not seem such a bad thing. But in reality, a warmer Earth would cause all sorts of problems. The ice caps at the north and south poles could begin to melt. Sea levels would rise, and low-lying countries, such as Bangladesh or the Netherlands, might disappear beneath the sea. Other effects, such as the spread of tropical disease, are very hard to predict—but they could happen.

Unfortunately, as with disasters that haven't happened, scientists cannot agree on whether global warming is a real threat to the earth or not.

Global warming would melt glaciers and icebergs, increasing sea levels all over the world.

Glossary

atmosphere blanket of gases that surrounds the earth

bacteria one of the simplest and smallest forms of life. They live in large numbers in the air, soil, and water.

compensation money paid to make amends for damages

contaminate to pollute a thing or place with poisonous chemicals

dredge to scoop up mud from a channel or river

environment external surroundings; the land, water, and air around us

enzymes special proteins that speed up the chemical changes necessary for life

evacuate to move people away from a dangerous place until the danger is over

fertilizer food for plants used by farmers to grow better crops

fumes unpleasant smoke or gas. Some fumes are toxic.

Greenpeace organization that campaigns to save the environment

negligence lack of proper care and attention

OECD (Organization for Economic Cooperation and Development) an organization of mostly North American and European countries, which works to promote the economic and social welfare of its members, as well as for people in developing countries

pesticide chemical used to kill or control insects on crops

pressure gauge instrument that shows the force of steam or air acting inside equipment

raw material natural material that has not yet been turned into a finished product

refinery place where oil is refined (impurities are removed)

sabotage damage done to something on purpose

synthetic describes a material that is made from chemicals

toxic poisonous

United Nations association of different countries that work together for international peace and security. One of its purposes is to aid refugees and victims of disaster, war, and poverty

valve small flap or door inside a pipe that can turn on or off like a faucet

World Health Organization agency of the United Nations established in 1946 to help prevent the spread of diseases worldwide

More Books to Read

Anderson, Cathy and Jeri Hayes, editors. *Chemicals: Choosing Wisely.* White Plains, N.Y.: Seymour, Dale Publications, 1997.

Diamond, Arthur. *The Bhopal Chemical Leak.* San Diego: Lucent Books, 1990.

Lampton, Christopher. *Chemical Accident.* Brookfield, Conn.: Millbrook Press, Inc., 1994.

Index